This SUPER AWESOME Book Belongs to:

Winner of the Art Competition: Charlotte Ray

Marine Biology

April Chloe Terrazas

Book 9 of the Super Smart Science Series™
Ages 0-100

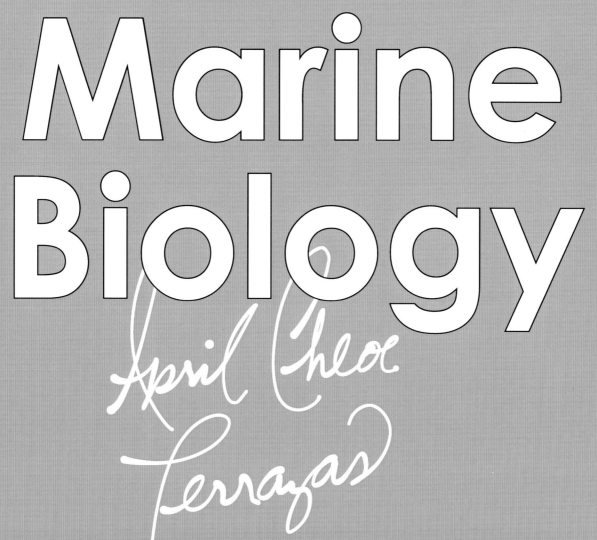

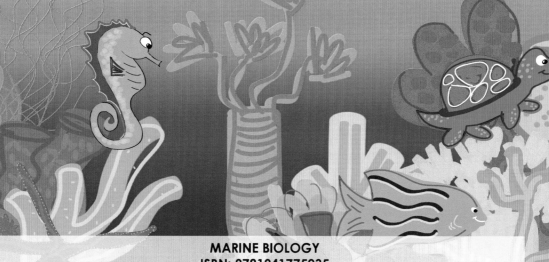

MARINE BIOLOGY
ISBN: 9781941775035
April Chloe Terrazas, BS University of Texas at Austin.
Copyright © 2014 Crazy Brainz, LLC

Visit us on the web! www.Crazy-Brainz.com

Cover design, illustrations and text by: April Chloe Terrazas

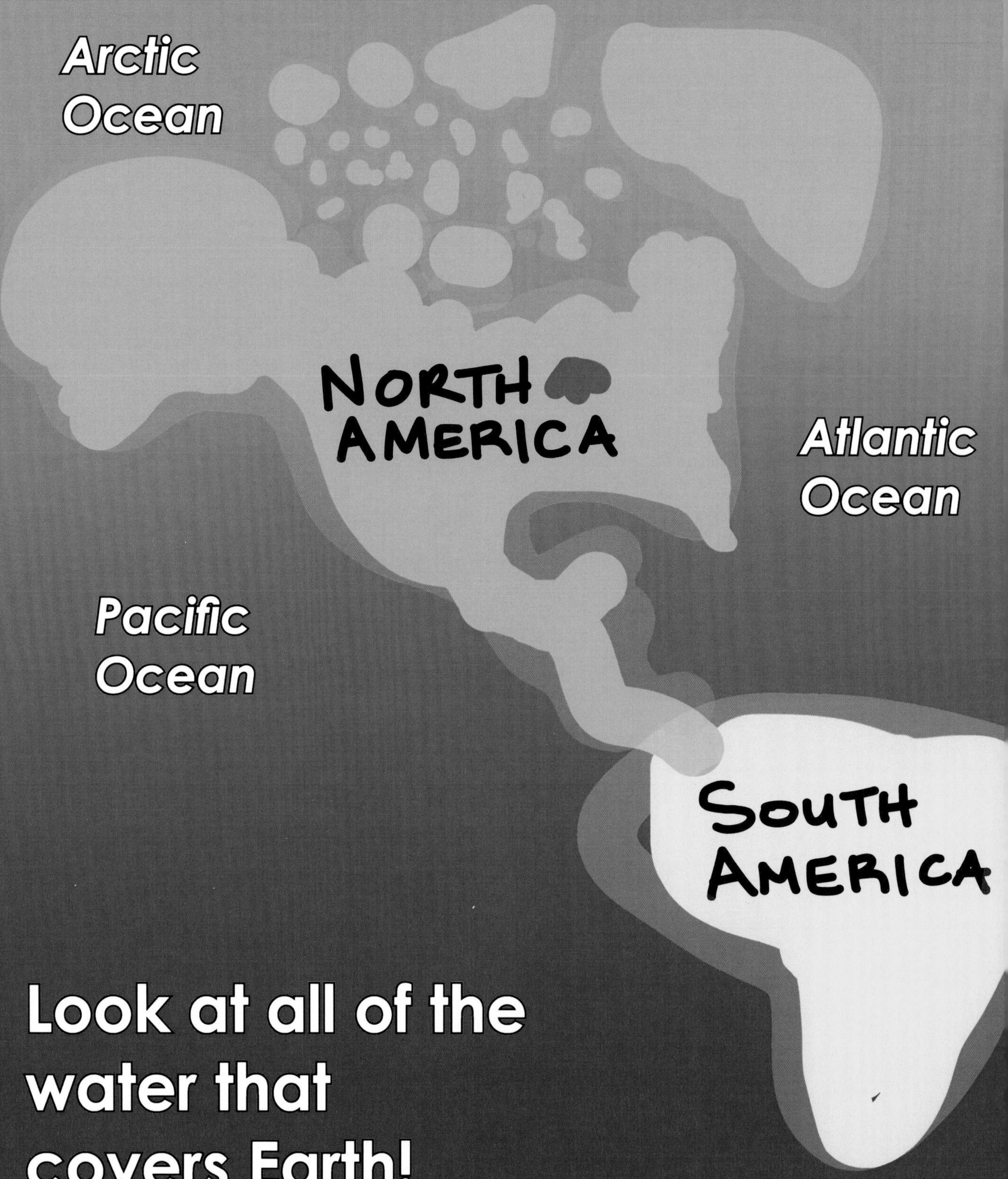

Look at all of the water that covers Earth!

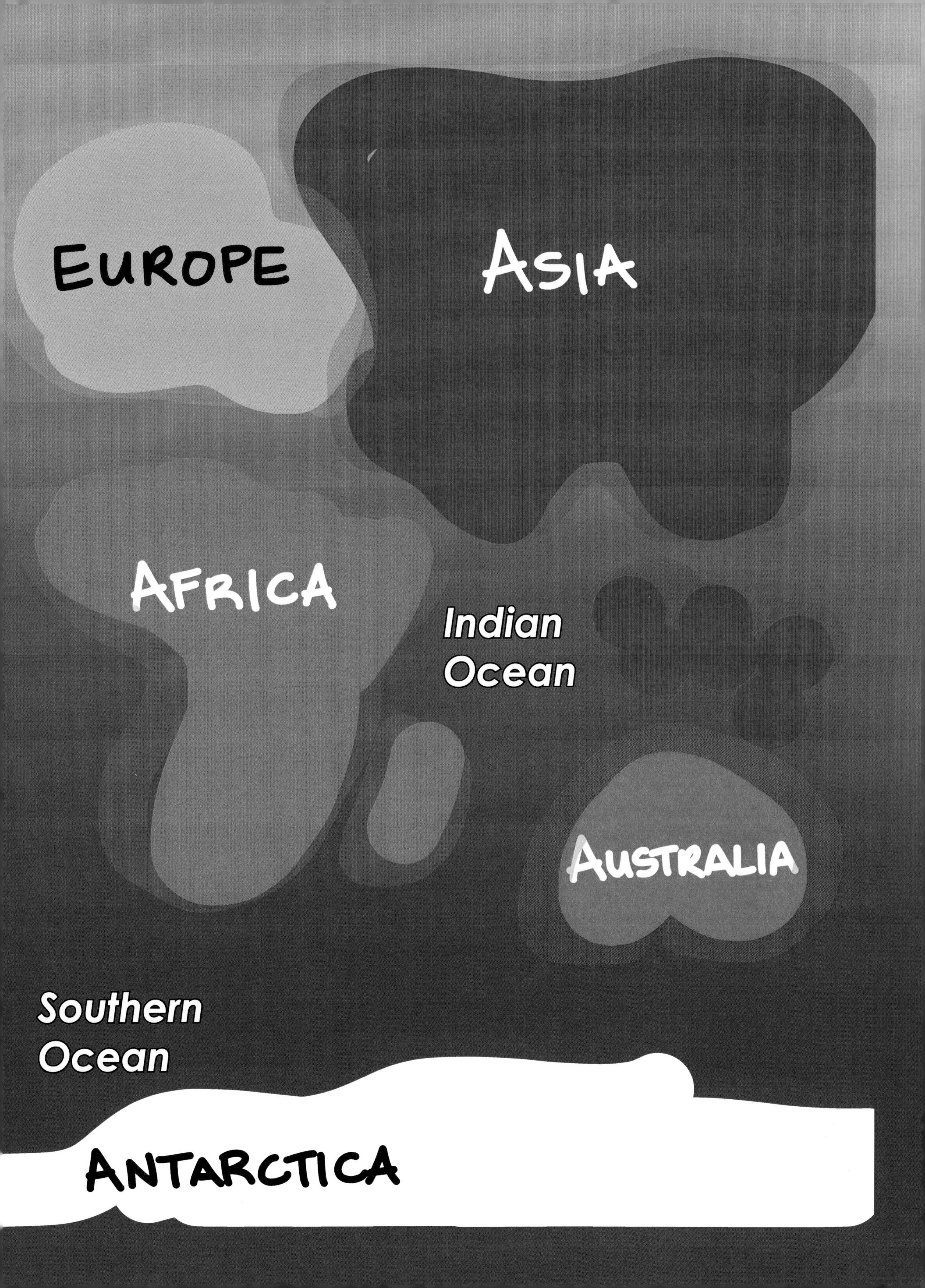
EUROPE
ASIA
AFRICA
Indian
Ocean
AUSTRALIA
Southern
Ocean
ANTARCTICA

Marine biology means "study of life in the sea."
(Marine = sea,
Biology = study of life)

Over 70% of Earth is covered by water. The oceans are very big, and very deep.

There are many forms of life in the vast oceans on Earth, we have only discovered some of them.

Some sea creatures are very small, some are very big.

You are becoming a
Marine Biologist!

You are going to learn about
sea water composition,
sea plants and sea animals.

Are you ready to begin?

Turn the page
and get started!

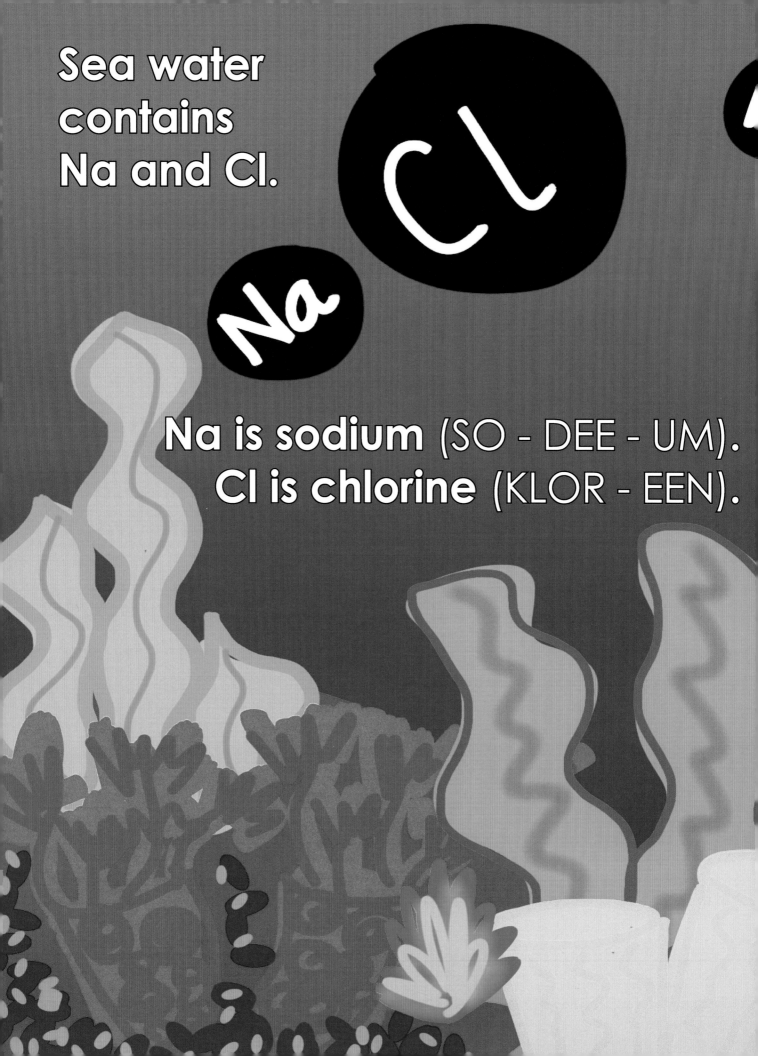

PHYTOPLANKTON
(FI - TOE - PLANK - TUN)

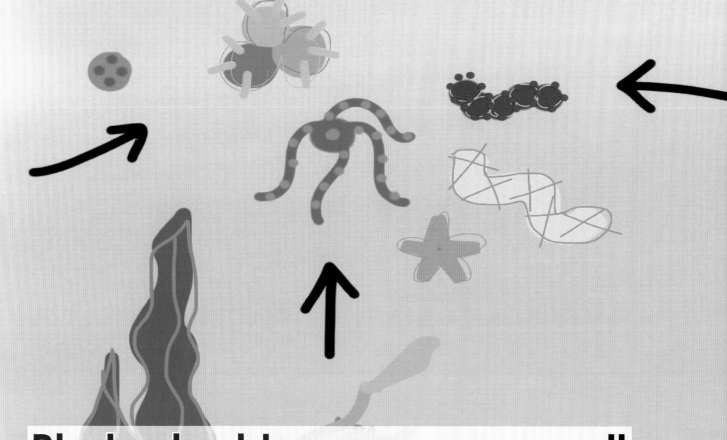

Phytoplankton are very small plants that serve as food for many sea creatures.

Phytoplankton are <u>VERY</u> small. They are just a tiny dot compared to an ant!

Phytoplankton require sunlight energy for photosynthesis.

PHYTO = plant

ZOOPLANKTON
(ZOO - PLANK - TUN)

Zooplankton are microscopic (very small) like phytoplankton.

(MI - KRO - SCOP - IK)

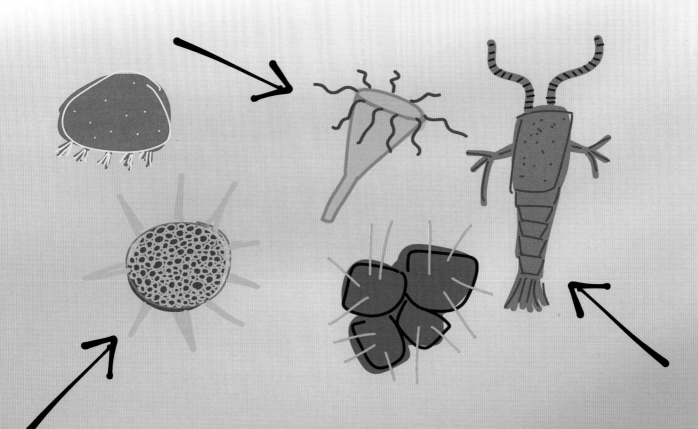

Zooplankton are mostly invertebrate animals.
(IN - VUR - TEH - BRUT)

Zooplankton provide food for a wide variety of marine life.

ZOO = animal

KELP (KELP)

Kelp is a plant.
Kelp provides food and
shelter for many sea creatures.

Kelp can grow up to
18 inches in ONE DAY!

Kelp relies on the sun to grow so it is near the surface of the water.

Kelp grows by photosynthesis.

What other sea plants grow by photosynthesis?

What is the definition of
"Marine Biology"?

What are the names of the
elements that make
sea water salty?

What is the name of the very
small plants that provide food
for many sea creatures?

What are the meaning of the
terms "phyto" and "zoo"?

What is the name of the very small invertebrate animals that provide food for many sea creatures?

How many inches can kelp grow in one day?

How do green plants, like kelp, grow?
(hint: it starts with a "P")

Fantastic!
Next, we will learn about animals that live in the ocean.

SPONGE (SPUNJ)

Sponges **are sessile.** (SES - EL)
This means they do not move.

Sponges eat by taking in water through the OSTIA (OS - TEE - UH) *(small holes on the side of the sponge)* **and feed on the phytoplankton and** zooplankton **in the water.**

The water and waste then exits the top of the sponge, called the OSCULUM (OS - Q - LUM).

Sponges like to live in warmer, tropical waters.

Sponges do not have a nervous system. A sponge will not respond if you touch it.

SEA STAR (C - STAR)

Sea stars **are regenerative.**
(RE - JEN - ER - UH - TIV)
This means they can grow a new body from just one piece of their arm!

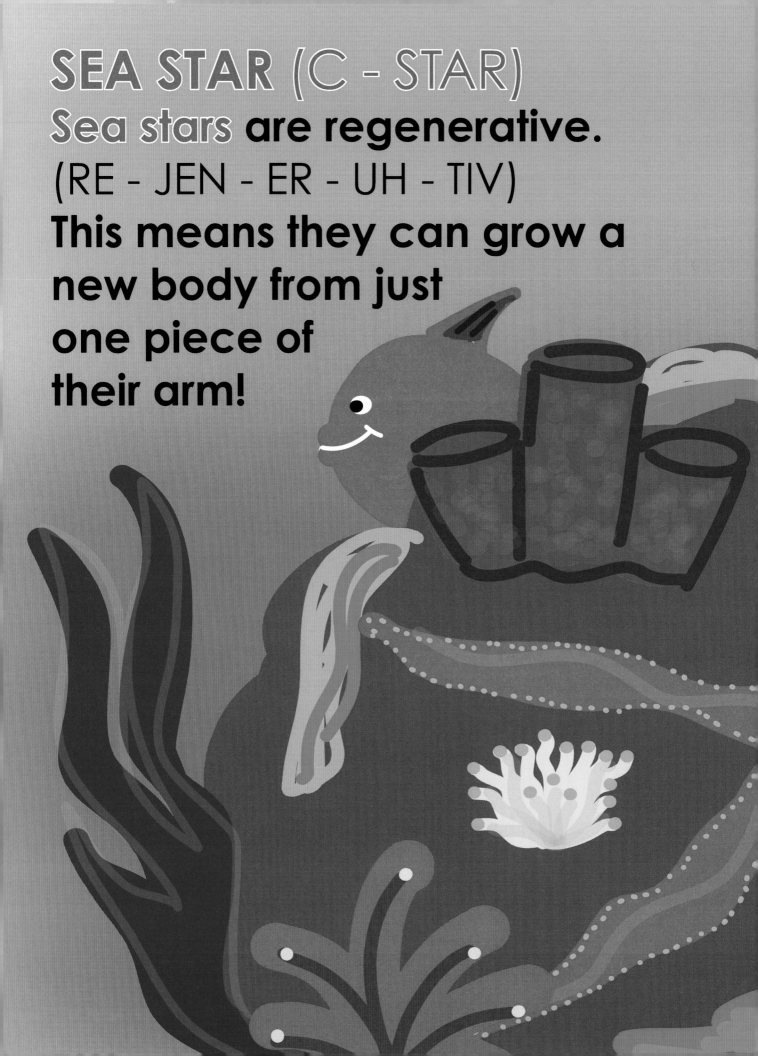

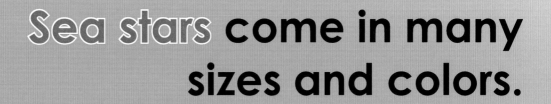

Sea stars come in many sizes and colors.

Sea stars have no brain and no blood. Instead of blood, they have sea water in their circulatory system.

What other marine life do you see here?

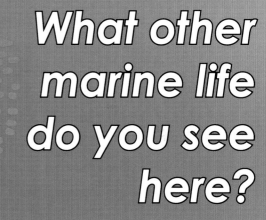

OCTOPUS (OK - TOE - PUS)

OCT = 8

Octopuses live on the sea floor in shallow to very deep waters.

The octopus has a very well developed nervous system which makes it very good at problem solving.

Octopuses secrete a dark ink to distract prey or as a defense to escape a predator.

How many legs does the octopus have?

JELLYFISH (JEL - EE - FISH)

Jellyfish have no bones and no brain. In fact, you can see right through them!

Jellyfish have sensors on their tentacles that detect movement. (TEN - TA - KLZ)

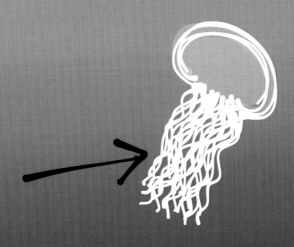

Jellyfish eat phytoplankton, invertebrates and small fish.

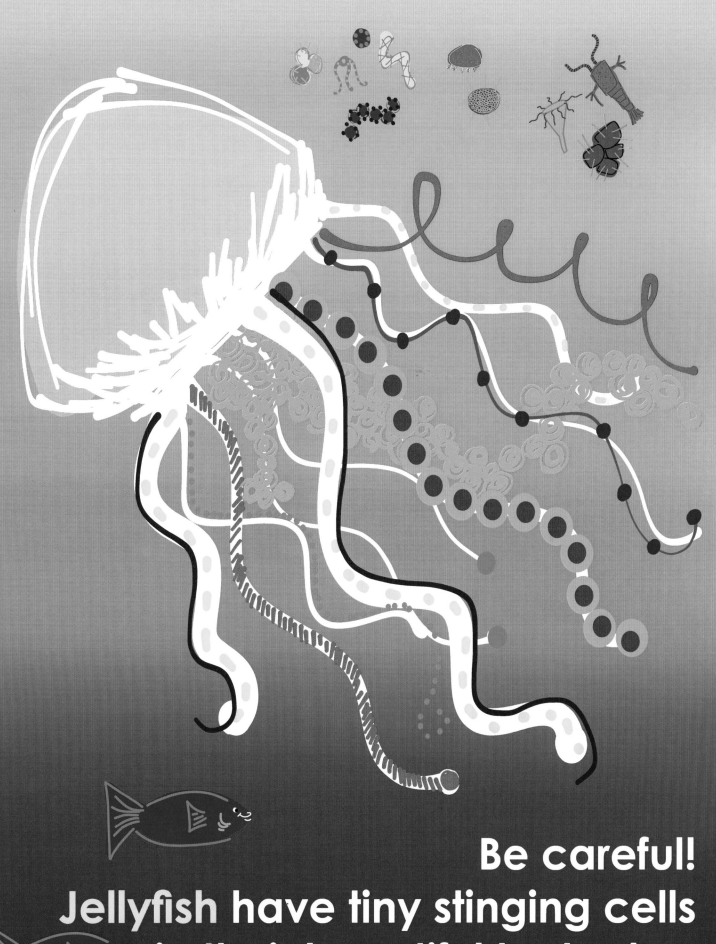

Be careful!
Jellyfish have tiny stinging cells in their beautiful tentacles.

Sponges do not move.
What is the term meaning
"does not move"?

What is the name of the small
holes on the sides of the sponge
that take in water and food?

What is the name of the large
hole at the top of the sponge
where water and waste exit?

Sea stars are regenerative.
What does this mean?

TRUE or FALSE

An octopus has 9 legs.

Octopuses secrete a dark ink to distract prey or as a defense against predators.

Octopuses have a poorly developed nervous system.

Jellyfish have no bones, or brain.

Jellyfish will not sting you.

Sea stars have blood in their circulatory system.

Sponges cannot sense touch.

SEAHORSE (C HORS)

Seahorses are very interesting little fish.

Seahorses can move their eyes independently. This means that one eye can be looking left and the other can be looking right!

Seahorses are always hungry. They can eat up to 3,000 pieces of food in just one day!

The seahorse is the only species where the male carries the babies.

Seahorses move by flapping their fins 30 to 70 times per second!

Sea Lion (C LI - UN)

Do you like to swim?

Sea lions **LOVE to swim!**

Sea lions **eat fish.**

Sea lions **live along coastlines so you can sometimes see them on the beach, laying in the sun.**

Sea lions **also hear and see very well, even under water!**

Stingray (STING - RAY)

Stingrays stay close to the ocean floor. They will bury themselves in the sand to hide from predators.

The eyes are on top of the body. The mouth is on the underbelly!

Stingrays are docile (DOSS - IL), which means they are friendly, and will swim near divers.

Be careful! Stingrays have a poisonous tail that can be harmful to humans.

Great White Shark
(GRAT WITE SHARK)

The **great white shark** is among the largest predatory animals in the ocean. It can grow between 15 to 20 feet long.

Great white sharks have up to 300 teeth!

Great white sharks can swim very fast, so fast that they can hurl their body completely out of the water!

Great white sharks can weigh up to 5000 pounds.

Coral Reef (KOR-UL REEF)

Corals **are animals!**
Some corals **are hard, making a rigid skeleton of calcium carbonate.** (CaCO$_3$, KAL - SEE - UM   KAR - BUN - ATE)

Other corals **are soft, moving back and forth with the flow of the water.**

Coral reefs can be very big.
It takes millions of corals to
make a coral reef.

Coral reefs provide shelter for
many sea creatures.

*What sea creatures do you see
hiding in the coral reef?*

How many pieces of food can a seahorse eat in one day?

Where do stingrays hide from predators?

Where is it possible to see a sea lion?

How much can a great white shark weigh?

What are corals?

Corals make a rigid skeleton of _____ _____.

TRUE or FALSE

Seahorses are the only species where the male carries the babies.

Sea lions do not like to swim.

Stingrays are friendly and will often swim near divers.

Great white sharks have only 30 teeth.

Coral reefs provide shelter for many sea creatures.

Na = Sodium, Cl = Chlorine
Phytoplankton
Zooplankton
Microscopic
Invertebrate
Kelp
Sponge
Sessile
Sea Star
Regenerative
Octopus
Jellyfish
Tentacles
Seahorse
Sea lion
Stingray
Docile
Great White Shark
Coral reef
$CaCO_3$ = Calcium Carbonate

You are now a Marine Biologist!
Next time you see the ocean,
think about all of the
amazing creatures inside!

Draw your favorite sea creatures!

CPSIA information can be obtained
at www.ICGtesting.com
Printed in the USA
BVXC01n1035290714
360784BV00003B/4